TALENT TO TRADE

BECOME A TECHNOLOGY WHIZ

Coding, Software Development & More

Kaitlyn Duling

Abdo & Daughters
MIDDLE GRADE NONFICTION
An imprint of Abdo Publishing
abdobooks.com

ABDOBOOKS.COM

Published by Abdo Publishing, a division of ABDO, PO Box 398166, Minneapolis, Minnesota 55439.

Printed in the United States of America, North Mankato, Minnesota
102024
012025

THIS BOOK CONTAINS RECYCLED MATERIALS

Design: Denise Hamernik, Mighty Media, Inc.
Production: Mighty Media, Inc.
Editor: Katherine Chu

Cover Photographs: Adobe Stock (cables on front cover, keyboard, network icon); Shutterstock Images (cables on back cover, circuit boards, cloud icon, computer background, monitor, mouse, security icon, servers, smart watch, USB)

Interior Photographs: Adobe Stock, pp. 14 (all), 15 (bottom), 50 (girl), 61 (top); AP Images, p. 55; HOLGER MUENCH/Flickr, p. 8; Mighty Media, Inc. (project photos), pp. 50, 51; Shutterstock Images, pp. 3, 4, 5 (all), 9 (right), 10 (all), 11 (all), 12 (all), 13 (all), 15 (top), 16, 17 (all), 18 (all), 19, 20, 21, 22, 23 (all), 24 (all), 25 (all), 26 (all), 27, 28 (all), 29, 30 (all), 31, 32 (all), 33, 34, 35 (all), 36 (all), 37 (all), 38 (all), 39 (all), 40, 41, 42 (all), 43 (all), 44 (all), 44–45 (background), 46 (all), 46–47 (background), 48 (all), 48–49 (background), 50–51 (background), 52, 53, 54, 56, 57, 58 (all), 59, 60, 61 (bottom); Smithsonian Institution/Flickr, p. 7; Science Museum/Wikimedia Commons, p. 6 (top); Michael Hicks/Wikimedia Commons, p. 6 (bottom); Paul Clarke/Wikimedia Commons, pp. 9 (left)

Design Elements: Adobe Stock (Polaroid frame, sticky notes, tacks); Shutterstock Images (circuit board texture, code texture, keyboard texture)

Library of Congress Control Number: 2024938322

PUBLISHER'S CATALOGING-IN-PUBLICATION DATA

Names: Duling, Kaitlyn, author.
Title: Become a technology whiz: coding, software development & more / by Kaitlyn Duling
Other Title: coding, software development & more
Description: Minneapolis, Minnesota : ABDO Publishing, 2025 | Series: Talent to trade | Includes online resources and index.
Identifiers: ISBN 9781098294984 (lib. bdg.) | ISBN 9798384915034 (ebook)
Subjects: LCSH: Computer programming--Juvenile literature. | Web site development--Software--Juvenile literature. | Design and technology--Juvenile literature. | Communication and technology--Juvenile literature. | Science, Applied--Juvenile literature. | Jobs--Juvenile literature. | Trades--Juvenile literature.
Classification: DDC 601.4--dc23

CONTENTS

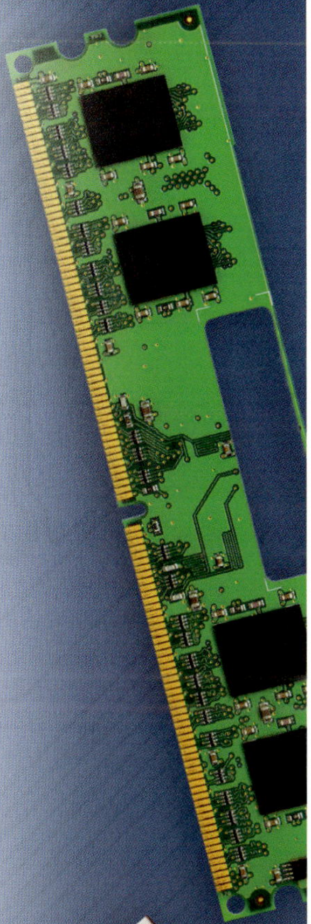

TALENT TO TRADE

Do you love computers? Have you ever wanted to design your own website? Have you thought about creating your own computer application or building a custom network? If your answer to any of these questions is yes, you might be the right fit for a job in computer and information technology. Professionals in this field create and support computer applications, programs, systems, and networks.

Becoming a technology whiz requires time, dedication, and hard work. Experts need to be comfortable spending hours at a desk working on a computer. They may need to learn about programming languages, network management, cloud computing, and more. Jobs in the technology industry combine many different areas of knowledge. These can be anything from computer hardware and software to Wi-Fi networks and computer security.

In this book, you'll learn about the history of computing and various jobs available in the industry. You will explore tools, skills, and techniques utilized by technology experts. You'll discover pathways to begin working toward your own career in computer and information technology. And finally, you'll learn some clever ways to turn your talents into a trade.

The Analytical Engine had four components. Data was entered into the reader. The store held the data, which the mill calculated. And the printer printed the results.

In 1963, Douglas Engelbart invented the computer mouse. It was carved out of wood and had only one button.

TECHNOLOGY THROUGH TIME

Computers may seem like a modern invention. But their history can be traced back hundreds of years! Some historians consider the modern computer's first ancestor to be the abacus. This is a mathematical tool created in ancient Mesopotamia.

The abacus and similar tools evolved over hundreds of years. In the mid-1830s, English inventor Charles Babbage started developing the Analytical Engine. It was meant to solve mathematical calculations. Many historians consider it to be the world's first computer. However, Babbage never completed his invention.

Fellow British mathematician Ada Lovelace saw promise in Babbage's machine. In the 1840s, she described how it could be programmed to solve equations. Her notes became the basis for the world's first computer program.

In the mid-twentieth century, the idea of the modern computer began to take shape. Throughout the 1930s, British mathematician Alan Turing worked on the theory of computing. He became an expert in cryptography. Then, he developed the Turing Machine. This simple computational device could analyze and solve mathematical equations. Turing later pioneered the field of artificial intelligence. Because of his many accomplishments, Turing became known as the Father of Modern Computing.

Computer scientist Grace Hopper (*center*) designed some of the earliest computer languages for the military and the private sector during World War II. She was later called the First Lady of Software.

With the 1940s came World War II, and advancements in technology helped the war effort. The world's first programmable, general-purpose, electronic, digital computer was completed in 1945. It was called the Electronic Numerical Integrator and Computer, or ENIAC.

The 1960s and 1970s saw key inventions in computing, such as the computer chip and mouse. In 1971, the floppy disk allowed data to be shared between computers. And in 1973, Ethernet connected multiple computers. Soon, companies began developing new software and smaller, more affordable hardware. Networks of connected computers were small. But they would soon connect people across the world.

In 1989, British scientist Tim Berners-Lee was working at the European Organization for Nuclear Research, known as CERN. He developed software that could automate information-sharing between universities around the world. Berners-Lee called it the World Wide Web (WWW). After CERN made the software available to the public, it took off. In 1990, students at McGill University in Montreal created the first web search engine, known as Archie.

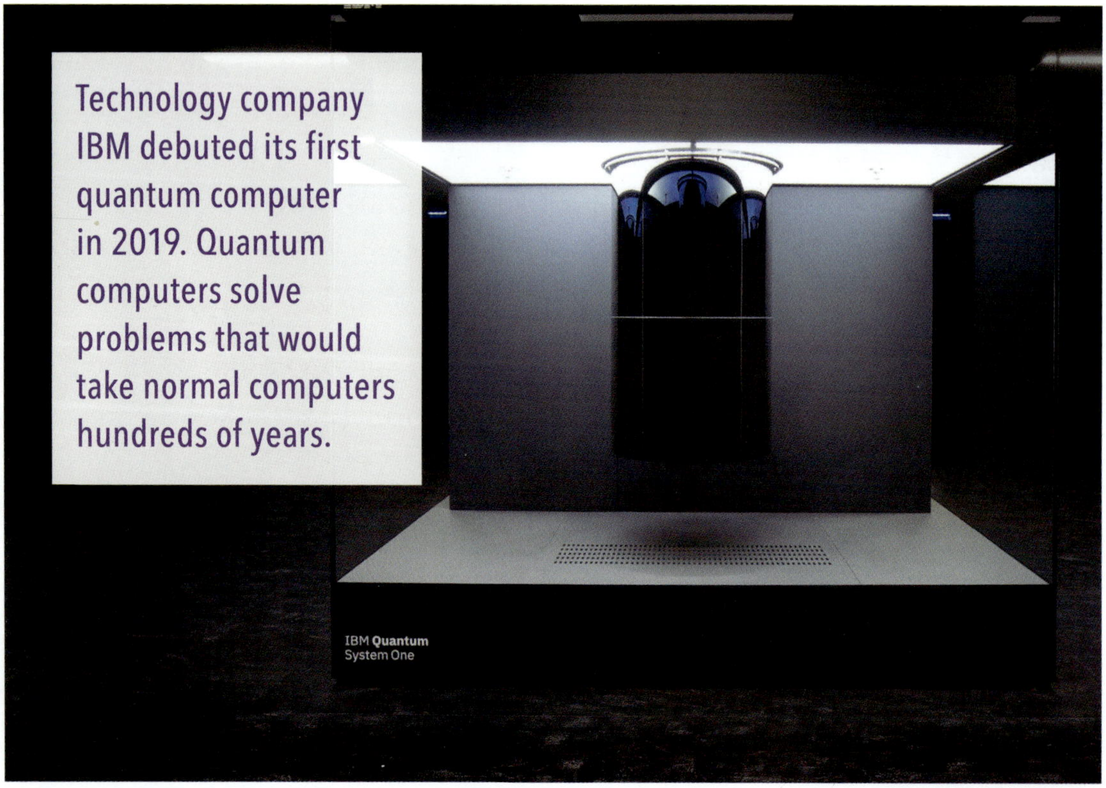

Technology company IBM debuted its first quantum computer in 2019. Quantum computers solve problems that would take normal computers hundreds of years.

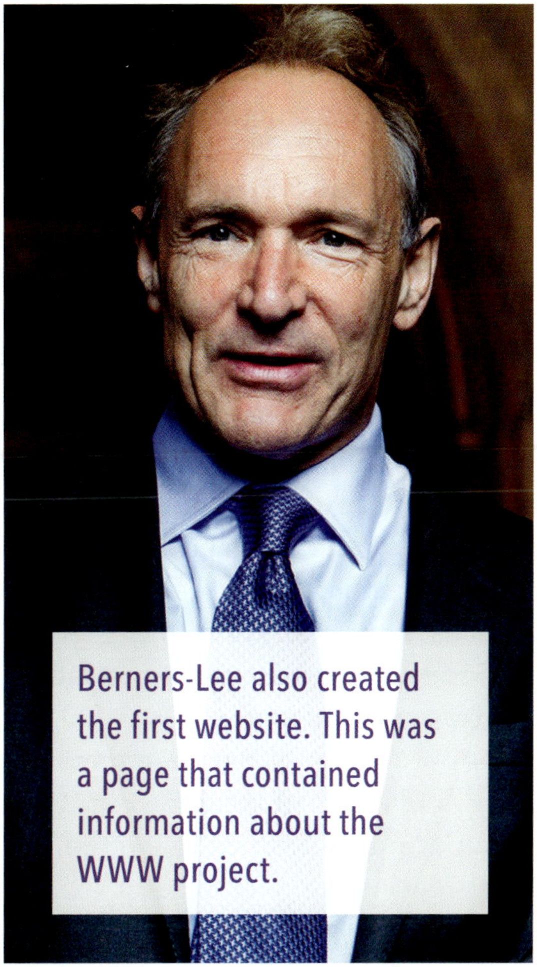

Berners-Lee also created the first website. This was a page that contained information about the WWW project.

The Google search engine was introduced in 1998.

In 1999, computer users were introduced to Wi-Fi. This allowed computer networks to exchange data wirelessly using radio waves. But with Wi-Fi came people who used the internet for criminal activity. So, network security experts were needed to keep networks protected. This newly connected world was full of opportunity—good and bad.

In the twenty-first century, the world of computer technology has only continued to grow. In 2022, OLCF-5 was created. It was the first public exascale supercomputer, also known as the world's fastest supercomputer. OLCF-5 is expected to keep watch over the US's nuclear weapons. It will also help solve complicated scientific problems—all in the blink of an eye.

Between 2022 and 2032, the computer and information technology industry is expected to grow by about 8 percent per year. In the following pages, you'll learn what it takes to become a technology whiz. You might even be inspired to work toward a career in computer and information technology!

TOOLS OF THE TRADE

Take a look at some of the tools tech professionals use to create and support computer networks, systems, and applications.

HARDWARE

The term *hardware* refers to the physical parts of a computer. All computers need hardware to work.

CASE

A computer's case is its outer shell. When you see a laptop or desktop computer, the case is the plastic or metal enclosure. This holds and supports most of the other hardware. It also protects them from physical and electrical damage.

KEYBOARD

Keyboards are used to input text into a computer. A keyboard's *keys* represent individual letters, numbers, symbols, and more. Computer programmers and coders work extensively with keyboards. This is because they spend much of their time writing computer code. Code is the instructions that a computer follows.

Some programmers and coders prefer split keyboards. These keyboards separate the keys into two or more sections. This helps the programmer or coder rest their hands in a more comfortable or natural position as they type.

MONITOR

A monitor is a device that connects to a computer. It displays images, text, and videos on a screen. Some computer experts use more than one monitor in their work. Others may prefer one especially large monitor. Web developers and digital designers work with visual elements. These include typography, graphics, and other aspects of a web interface. For these experts, having a computer monitor that accurately and clearly displays visual elements is important.

MOTHERBOARD

A motherboard is a rigid circuit board found inside every computer. It's usually made of fiberglass and copper. The motherboard is the main connection and communication point for the computer's other electrical hardware components. These include the central processing unit (CPU) and random access memory (RAM).

CENTRAL PROCESSING UNIT

Computers have many processors. These electric circuit boards give instructions to a computer. The CPU is the computer's most important processor. Located on the motherboard, it's like the computer's brain. The CPU controls everything the computer does. This includes the execution of the most basic instructions that run the computer. The CPU also processes inputs, stores data, and controls outputs.

RANDOM ACCESS MEMORY

RAM is a type of hardware usually attached to the motherboard. It is similar to short-term memory in our brains. It temporarily stores data. It can also access data at high speeds. RAM stores data currently being used by the computer's CPU. This can be the data needed to access the internet through a web browser, open a word processing program, and more.

GRAPHICS PROCESSING UNIT

A graphics processing unit (GPU) is an electronic circuit. Like a CPU, it processes inputs, stores data, and controls outputs. What makes a GPU different is the type of data it processes. A GPU handles graphics, images, videos, and visual effects. It also allows images and videos to display on a computer monitor.

NETWORK HARDWARE

Network hardware includes all the physical parts used for a network. These parts are required for communication between different computers and other digital devices.

MODEM & ROUTER

Computer professionals who maintain or secure networks regularly work with modems and routers. Modems are devices that connect to an internet service provider (ISP). They send and receive signals from the ISP. Most modems physically connect to the ISP through a cable. They bring internet service into a home, an office, or another space.

A router is a piece of hardware that connects a modem to other

devices on a small network. These can be smartphones, smart TVs, laptops, tablets, and more. Routers help these devices communicate with each other and connect to an ISP. Many home modems contain a Wi-Fi router. Devices can connect wirelessly to the modem and router using Wi-Fi. They can also connect through a wired connection such as a cable.

NETWORK SWITCH

While a router connects devices to the internet, a network switch only connects wired devices together within a network. These switches can connect hardware such as computers, printers, and servers. Network administrators use network switches in large offices and data centers with hundreds of computers. For this reason, information security analysts may use network switches to keep networks secure.

ACCESS POINT

An access point is also known as a wireless access point, or WAP. It is a piece of networking hardware that allows multiple wireless devices to connect to a network. WAPs can provide complete coverage for wireless devices within a large office or other space. They work by minimizing "dead spaces," or areas where the Wi-Fi signal might be low or nonexistent. Network administrators may use WAPs when building and maintaining networks.

REPEATER

Similar to a WAP, a repeater is a device used to expand a network's reach. It takes in a signal and retransmits it with higher power. A repeater can retransmit a stronger signal over a wider area than a WAP can. It can also transmit through obstructions, such as walls. A repeater is also a secure form of data transmission. This is especially important for security analysts as they keep networks protected.

SOFTWARE

The term *software* refers to all the nonphysical components of a computer. These include programs, data, and other information used by the computer. Software is made up of multiple programs. Each program provides instructions that help tell a computer what to do.

OPERATING SYSTEM

The most important software on a computer is its operating system (OS). The OS manages the computer's hardware, software, memory, storage, and processes. It also allows a computer user to communicate with the computer. If a person is using a computer, they are using an OS. Most computers come with a preloaded OS. However, users can change a computer's OS or build their own. A computer systems administrator might choose the same OS for everyone in an office or an entire company. Windows is one example of an OS.

FIRMWARE

Firmware is a type of software that provides basic instructions to certain types of hardware. This allows digital devices to start up, perform basic tasks, and communicate with other devices. Cameras, smartphones, routers, scanners, and other devices need firmware to run effectively. Software developers create, test, and improve firmware. And computer systems administrators often update any firmware their companies' devices use.

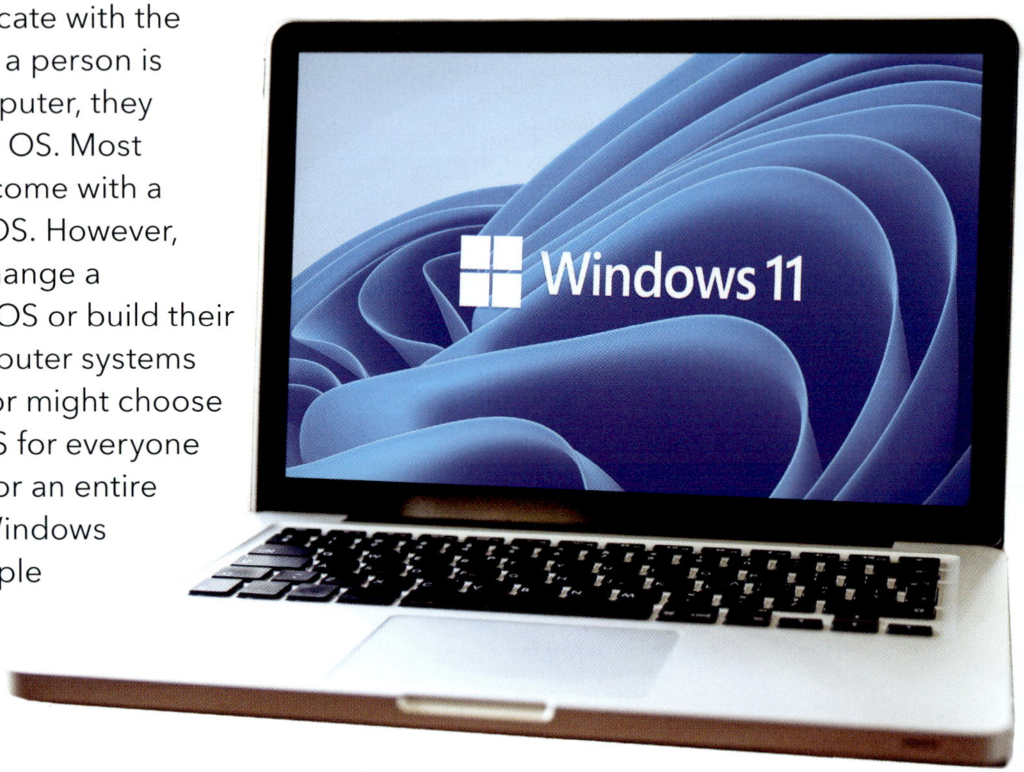

APPLICATION SOFTWARE

Application software is a type of computer program. It performs specific functions and allows a computer user to complete certain tasks. It does this by instructing a computer on what to do or how to complete a task. There are countless different types of application software available for computers and other digital devices. Examples include web browsers, messaging apps, game apps, and more. Software developers create, test, and improve application software. And many computer programmers and coders write the code that makes up application software.

MIDDLEWARE

Usually, application software connects and communicates directly with a computer's OS. But sometimes, middleware acts as a bridge that connects the two. This enables better communication and data management. Software developers work to create, test, and improve middleware.

WEB BROWSER

A web browser is a type of software application. It allows users to find, access, and view websites on the internet. Software developers plan and design how web browsers work. Computer programmers and coders write the code that turns a software developer's idea for a web browser into reality. And web developers and digital designers create and maintain the websites that users access through web browsers.

FIREWALL

A firewall is a network security system that monitors all internet traffic. Internet traffic is the amount of data moving in, out, and through the network. Firewalls can also restrict internet traffic within a network. Information security analysts and network and computer systems administrators work directly with firewalls to provide protection and security to networks.

SPECIAL SKILLS

Explore some of the skills that technology experts need to do their jobs.

ATTENTION TO DETAIL

Computers are intricate devices made up of countless parts. Once loaded with software and connected to a Wi-Fi network, they become infinitely more complex. This is why technology professionals must have expert-level attention to detail. A small mistake in a line of code can cause an OS or other software to crash or stop working properly. And overlooking a potential security threat can put an entire network at risk.

CLOUD COMPUTING

Cloud computing is the ability to store and access computing services and resources on the internet, or "the cloud." These services can include servers, storage, databases, software, and more. The cloud eliminates the need for certain forms of computer hardware. As more services are offered though the cloud, technology professionals need to better understand cloud computing. Information security analysts and network and computer systems administrators need to be especially knowledgeable about cloud security.

COLLABORATION

Many roles in computer and information technology are extremely collaborative. Large teams often create computers and software. Developers, programmers, and coders are part of that larger team. Likewise, network and computer systems administrators and information security analysts may work for companies or organizations with numerous employees. They need to work collaboratively to make decisions and solve problems.

CRYPTOGRAPHY

Information security analysts and other technology experts may work with sensitive information. Those who do must be competent and stay updated on cryptography tools and procedures. These tools add an extra layer of protection to ensure privacy.

NETWORK MANAGEMENT

Network and computer systems administrators are responsible for managing a network of computers and other devices. They must be skilled and experienced at managing a network's hardware, software, and Wi-Fi. This can include taking inventory and updating software. It may also involve setting up and maintaining any hardware necessary for seamless Wi-Fi coverage.

PROGRAMMING LANGUAGES

Computers understand binary code, also known as machine language. This simple language consists of ones and zeros. But it is very difficult for humans to

understand. So, programmers and coders usually use programming languages. These languages are translated into machine code for the computer to understand.

Programming languages are made of special vocabularies and grammar rules. These help give instructions to a computer. The programming language a professional uses depends on what kind of task they want a computer to perform. For example, web developers use languages such as JavaScript, Hypertext Markup Language (HTML), and Cascading Style Sheets (CSS) to create websites. Software developers, programmers, and coders use languages such as Python, C, C++, Ruby, and Structured Query Language (SQL).

These languages help experts create different types of software. Most professionals master at least one language. But a technology whiz may use multiple languages over the course of their career.

TESTING & TROUBLESHOOTING

Whether creating a website, coding an app, or securing a network, an expert needs to know how to test their project. They do this to ensure everything is working the way it's

supposed to. During the testing process, they troubleshoot any issues that come up. Programmers and developers may go through several rounds of testing and troubleshooting before they have a finished product.

WORKING UNDER PRESSURE

Some technology experts work in high-pressure positions. Information security analysts respond to threats that can adversely affect entire companies. They must work with speed, confidence, and confidentiality. Developers and programmers sometimes need to create software under pressure. This may happen if a company is trying to quickly produce a product. And network and computer systems administrators can face mounting pressure when diagnosing and solving problems with large computer networks and systems.

CAREERS IN COMPUTER & INFORMATION TECHNOLOGY

Technology experts can choose from a variety of specialties on which to focus their studies and skill-building. Let's explore some career paths for those interested in tech.

WEB DEVELOPER & DIGITAL DESIGNER

Have you ever used a website? If so, you've seen and used the work of web developers and digital designers. Web developers are responsible for creating and maintaining websites. They handle technical details such as the website's speed and how much traffic, or visitors, it can handle. Digital designers control how a website looks. They also create the website's layout and functions. When doing this, they consider the user's overall experience when navigating a website.

CLIENT CONSULTATION

Web developers and digital designers may work for a company or an individual. Before creating a website, developers and designers determine their client's needs. A client may want a specific design, such as fun graphics for young children. They may also ask for a

particular functionality, such as online sales. Client consultations help define the overall vision for the project as well as the client's specific needs.

CREATE WEBSITES

After the initial consultation, web developers and digital designers begin planning and designing the website. The two often work together to settle on specific designs. In addition to web pages, they may need to create technical aspects. These can be email services, databases, graphic displays, sales functionality, and more.

Most developers use a programming language to create a website. Developers with little or no coding experience may use a website builder. Designers may also use this online tool to create the look of a website. And some developers might use a content management system. This is a web application or program used to create and manage websites.

CODING WEBSITES

There are various programming languages web developers use to create websites. HTML tells a browser how to show the website's words and images. XML allows programmers to store and share data on a website. JavaScript creates interactive websites. And CSS creates the design of a website.

Developers may use multiple languages together when creating a website. They also use different languages for the front end and the back end of the website. The front end is the part of the website that users see. Front-end coding allows users to interact with the page. This happens when a user plays a video, highlights text, expands images, and more. The back end of a website is everything a user doesn't see. This includes the databases, servers, and other elements that make websites functional.

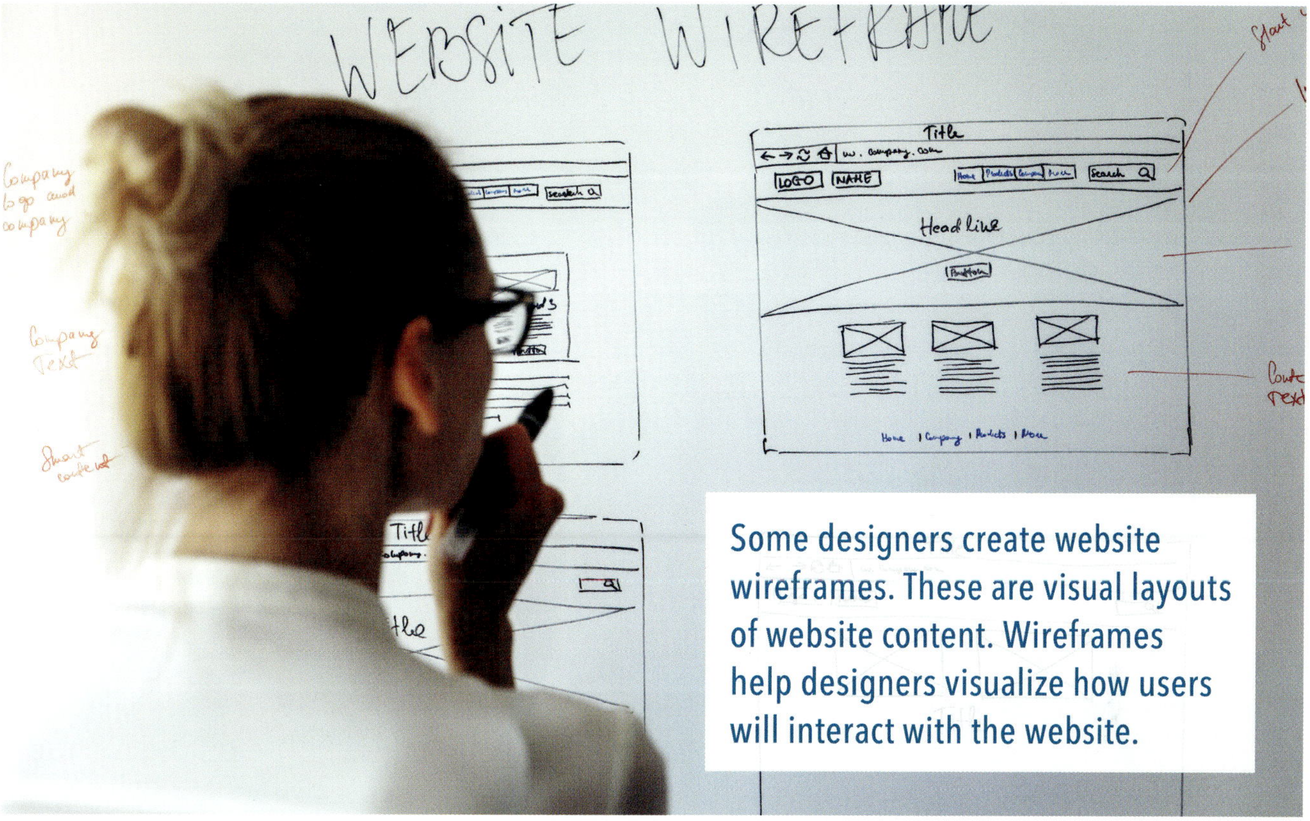

Some designers create website wireframes. These are visual layouts of website content. Wireframes help designers visualize how users will interact with the website.

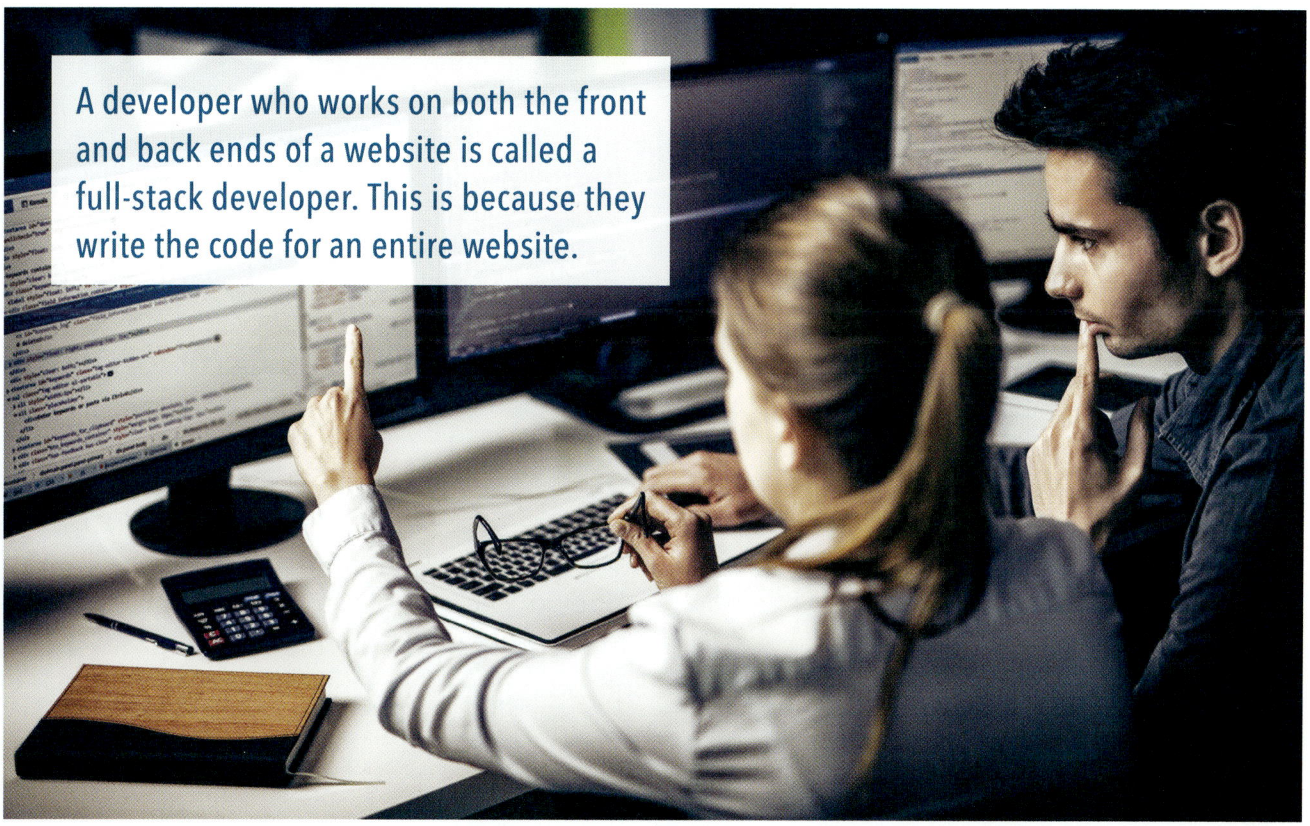

A developer who works on both the front and back ends of a website is called a full-stack developer. This is because they write the code for an entire website.

Most website mock-ups show what a website will look like on different devices. These can include computer monitors, tablets, and smartphones.

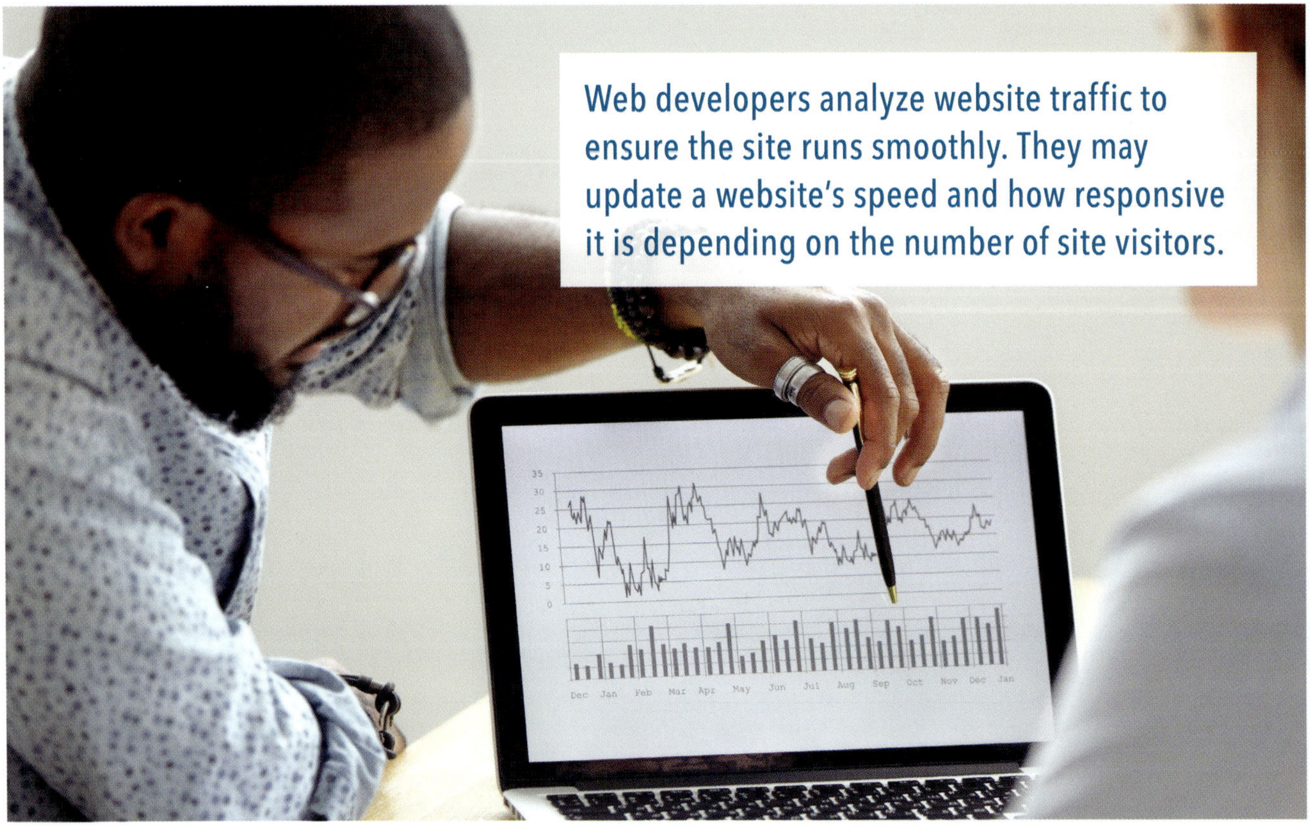

Web developers analyze website traffic to ensure the site runs smoothly. They may update a website's speed and how responsive it is depending on the number of site visitors.

CREATE MOCK-UPS

Before a website is published, web developers and digital designers create a mock-up. This is a sample version they share with the client. It allows the client to see what a website will look like and understand how it will function. Once the client approves the website, it is published, or made accessible on the internet.

WEBSITE MAINTENANCE

Web developers do more than design, create, and publish websites. They may also continue working with clients to maintain their websites. They provide updates, analyze website traffic, test for errors, troubleshoot issues, and more.

INTEGRATE MEDIA

Digital designers focus on the layout and design of a website. Some websites are entirely text-based. But most include media such as graphics, video, and audio. Designers can create custom graphics, video players, and more. Then they work with web developers to integrate their media choices into the website's code. For example, a website with lots of fast-moving video needs to have the capacity to handle it. Some designers and developers may specialize in creating media for a website. Others might work on all aspects of a website.

SOFTWARE DEVELOPER

Software developers design, build, manage, and apply software applications, programs, and systems. They work on everything from video games and smartphone apps to OSs and word processors.

DESIGN & DEVELOP SOFTWARE

Software developers create software to do specific tasks. There are many different types of software developers. Some work to update or change software created by other developers. And some create new types of software. For example, an applications software developer may create or update software that performs a user-focused task. This can be building word processing software. Word processing software helps computer users create, edit, save, and print word documents. An example of word processing software is Microsoft Word.

Other developers may work on larger systems or networks. They may create software to support or manage different technologies or devices. And many systems software developers design and develop OSs for businesses or organizations. They also create the software that keeps all digital devices and networks working smoothly and efficiently.

MODELS & DIAGRAMS

Some software developers write their own computer code. Some may even program the software for an entire application or system. But many software developers work only to design and plan the software they want to create. These developers work alongside programmers and coders. To communicate their designs and ideas, developers typically create models and diagrams

Many developers work either alone or in small groups depending on the type and size of the projects they are working on.

Developers may use a variety of model or diagram styles or languages. They choose the style based on who they are communicating with and what they are trying to show.

Many software developers include a README text file with their software. This text file may include detailed instructions, licensing information, any known issues, who created the software, and more.

for programmers to follow. Programmers and coders then write the code for the programs to work. This means developers need to be well-versed in giving clear instructions. They also manage the software building process from start to finish. They do this to ensure it follows the client's specifications.

MAINTENANCE, TESTING & UPGRADES

Once software has been created, it needs to be tested and adjusted until it meets a client's needs. After releasing the software to the client, it may still need ongoing maintenance and upgrades. Some software developers provide this maintenance. This usually includes fixing errors, addressing any difficulties users face, and more. Developers may also offer additional upgrades or make upgrade suggestions.

DOCUMENTATION

During and after the development process, software developers often create documentation for the client. This includes operational manuals, system instructions, and more. Developers also document any problems encountered while testing, during maintenance, or after upgrading the software. This helps guide other developers or programmers who work on the software after its release.

COMPUTER PROGRAMMER & CODER

The two titles–programmer and coder–are often used interchangeably, but they have some differences. The most distinct difference is that computer programmers plan, code, and test programs. Computer coders only write the code needed for a program. They do not plan or design software. Some coders know multiple programming languages, but most know only one.

WRITE PROGRAMS

Programmers work closely with software and web developers. They turn a developer's designs into working software programs that computers can follow.

Programmers must correctly interpret a developer's designs when building software programs. They need to plan and create the software to provide a specific result. Then they compile or write the code for the program or delegate these tasks to a coder.

UPDATE & TEST PROGRAMS

Computer programmers sometimes fix program errors, update programs, and even expand existing programs. They also run tests to make sure programs are functioning properly and delivering the promised output.

The responsibilities of a programmer and a developer sometimes overlap. For example, some programmers also design software.

Python, JavaScript, and C++ are some of the most used programming languages.

Some programmers may use parts of code found in code libraries. They may modify the existing lines of code and add their own code to create a custom program.

USE CODE LIBRARIES

A code library is a digital collection of pre-written, reusable code. Each library has a primary language, or code. Programmers use code libraries to perform specific tasks. These can be everything from generating sound effects or creating images to connecting computers or compiling data sets. For example, code from a library might help a programmer efficiently create a user login field. Code libraries help programmers and coders complete tasks faster than they would if they wrote the code from scratch.

WRITING CODE

Computer programmers and coders can't rely completely on code libraries. The programs they create may have similar parts, such as user logins. But programmers customize programs to fit their clients' needs. This means programmers must create new lines of code. To do so, they must understand one or more programming languages.

NETWORK & COMPUTER SYSTEMS ADMINISTRATOR

Network and computer systems administrators install, configure, and maintain computer networks and systems. They work for schools, financial institutions, governments, and more. Almost anywhere there is a computer network or system, you will find one or more of these administrators. Some oversee both networks and systems. But many specialize in either networks or systems. Network administrators work on all aspects of a network that connects the computers of an organization or business. System administrators work on all computer system software and hardware.

DETERMINE NETWORK & SYSTEM NEEDS

When an administrator starts a new position or a company opens an office, they will first determine the organization's needs. This can involve installing or updating hardware and software. It may also mean building out a network to ensure secure internet access for the organization. There are two typical network sizes:

- A local area network (LAN) connects users and applications that are close together. A LAN is usually used within one building.
- A wide area network (WAN) connects users and applications across a wide geographical area. This can be a state, a country, or even across the globe.

MANAGE HARDWARE

Once they have determined a company's needs, an administrator may suggest specific hardware for the company to purchase. System administrators manage and maintain servers, computers, and mobile equipment. And network administrators manage and maintain all network hardware. This includes routers, switches, and cables.

Smaller companies may hire one administrator to handle all network and system needs. Larger companies may have multiple administrators. Each handles a different part of the network and system.

Administrators may travel to different company sites to set up or fix computer hardware.

System administrators regularly create backup files. This way, if anything happens to a user's system, they can go back to the previously saved version.

Administrators may train users on how to upload and download data so a system stays secure.

MANAGE SOFTWARE

Computer systems administrators are typically responsible for setting up and maintaining an organization's software. This may include the OS, email software, data storage, and more. They also make needed upgrades and enable user access for company employees and others.

MAINTAIN NETWORK & SYSTEM SECURITY

Many network and computer systems administrators manage network and computer security systems. This is one of their most important tasks. They also add and train new users, assign security permissions, and resolve any security issues. All organizations and businesses need to keep their networks uncompromised. So, administrators work hard to secure user data, private information, financial records, and more.

EVALUATE NETWORK & SYSTEM PERFORMANCE

An administrator needs to stay updated on the latest system and network technology. This ensures that their organization's network is functioning efficiently and securely. It also ensures that they have the right software to best serve the organization.

DIAGNOSE & RESOLVE PROBLEMS

When problems arise with computer hardware and software, system administrators must respond promptly. They need to efficiently troubleshoot and fix any problems users may have. Network administrators also respond and solve network problems quickly. This is important so issues do not halt operations or compromise an organization's security.

INFORMATION SECURITY ANALYST

Planning and maintaining the security of an organization's network and systems is one aspect of a network and computer systems administrator's job. It's also the entire job of an information security analyst. Professionally certified information security analysts are experts when it comes to protecting networks.

USE & MAINTAIN SOFTWARE

To protect sensitive data, information security analysts use and maintain a variety of software. This includes firewalls, data encryption programs, and assessment tools. This software helps analysts evaluate devices and networks, identify potential threats, protect against viruses, and ensure privacy.

DISASTER RECOVERY PLANS

Every network and system needs a disaster recovery plan. Security analysts make, maintain, and test recovery plans. Information technology employees follow these procedures if a security emergency occurs within an organization's network. A recovery plan allows for continued operation of the information technology department and the organization. It can include everything from preventative measures to recovery steps to take following an emergency.

Many security analysts work with network and system administrators to improve network and system efficiency and security.

Security analysts keep track of how a company's system performs. Any small changes can indicate a security breach.

Analysts often test their companies' security programs to make sure they work properly.

INVESTIGATE SECURITY BREACHES

Even if an analyst is the best in their field, security breaches can still happen. Cybercriminals work hard to find vulnerabilities within security systems. An analyst identifies and stops any attempted cyberattack or other security emergency. Then they work to investigate what caused it. This helps them plan how to stop similar attacks from occurring in the future.

HELPING COMPUTER USERS

Information security analysts also work with computer users daily. They remind users to install new security software. They help users troubleshoot issues with that software. Analysts may also provide ongoing education on security. This includes how to safely access an organization's network, best practices for password encryption, and other helpful security information.

SPARK YOUR VISION

It's time to get creative! What do you love about computers and information technology? If you could create your own website, video game, or mobile app, what would it look like? Create a vision board that reflects these ideas. Let it motivate you to turn your talents into your trade!

Put your vision board where you'll see it on a regular basis, such as in your locker or next to your bed.

"Machines take me by surprise with great frequency."
—Alan Turing

Display sample templates, quotes, and photos that inspire your website or app design.

"The computer was born to solve problems that did not exist before."
—Bill Gates

EXPAND YOUR KNOWLEDGE

You've got the passion, vision, and motivation. You may not be ready to start formal training. But you can still take steps to learn about computers and programming languages on your own.

Use code blocks to learn simple programming with Kano or Scratch. Or check out programming websites such as Codecademy.com, Sololearn.com, or Python.org.

Take computer-related classes at your school or local library.

If you have Wi-Fi, computers, mobile devices, or other technology at home, ask your parents to show you how the home network works.

UPGRADE YOUR EXPERIENCE

Now that you've learned more about computers and software, you can start using your skills! Gain confidence as a tech whiz by testing your knowledge with hands-on experience.

Go on a "hardware field trip" at your local technology recycling center or repair shop. Use online guides to examine and understand any hardware you take home.

Seek out Capture the Flag (CTF) competitions in your area. These events will give you knowledge and experience in information security.

Offer to help friends and family members with their IT issues.

IGNITE YOUR CAREER

Ready to get started? Brainstorm some ways to apply your talents to make money. Someday, your side hustle may grow into a career!

Once you have some experience, volunteer as a leader for a coding, robotics, or computer club at your school or local library.

Offer to design websites for local nonprofits, community groups, or individuals.

Apply for an internship in computer and information technology.

SHARE AN E-CARD

Learn coding basics by using Scratch code blocks to create your own audio GIF e-card. Email or message your e-card link to family and friends!

SUPPLIES

smartphone or camera

computer

STEPS

1 Record a video with a smartphone or camera. Transfer the video file to your computer.

2 Use an online converter, such as Ezgif, to convert the video into a GIF. Save the GIF to your computer.

3 Go to the website https://scratch.mit.edu/projects/editor. If you don't have a Scratch account, create one and sign in.

4 In the bottom right of the screen, you'll see a Scratch Cat icon. Click on the trash bin on the top right of the icon to remove it.

5 In the same box as the icon, you'll see a circle button with a cat head and plus sign. When you hover your cursor over the button, a vertical menu will appear. Click on the top option, "Upload Sprite." Find and select the GIF file you saved in step 2.

6 To the left of the screen, you'll see colored blocks. Each color represents a different function that provides instructions for the GIF to follow. For example, the Events blocks tell the GIF what to do when something happens, such as a user clicking a button. The middle

box is your workspace. Drag blocks onto your workspace and connect them to create sets. Click on the green flag button at the top of the screen to preview your GIF.

7 Follow the example above to start a new block set that controls your GIF. Experiment by connecting different Motion, Looks, and Control blocks to your block set. Click on the green flag to preview your GIF!

8 To add sound, click on the Sounds tab on the top left of the screen. At the bottom left of the screen, you'll see a circle button with a speaker and plus sign. Hover your cursor over the button to view the vertical menu. Click on the second "Record" option.

9 If a pop-up appears asking permission for the web browser to access your microphone, click "Allow."

10 Click on the "Record" button to record a voice message. When you're done, click on the "Stop recording" button. Then click on the "Save" button.

11 Click on the Code tab on the top left of the screen. Connect the "broadcast 'message1'" Events block to the "forever" Control block's gap so it's above the other blocks in the gap.

12 Drag the "when I receive 'message1'" Events block onto your workspace to start a separate block set that controls the sound. Connect the "play sound 'recording1' until done" Sound block to the "when I receive 'message1'" block. Experiment by connecting different Control blocks to the second block set. Click on the green flag to see how they affect the sound!

13 When you're done, click on the "See Project Page" button in the purple bar at the top of your screen. Type your e-card's title into the top text box. Type a message into the Instructions box. Then, click the "Share" button. This will make your e-card available to anyone who has your e-card's link. If needed, follow the instructions to verify your e-card through your email.

14 Click on the "Copy Link" button on the bottom right of your screen. A box will appear where you can copy your e-card's link to share it with others!

Tip! Customize different e-cards for your family and friends to make their day special!

BECOMING A TECHNOLOGY WHIZ

EDUCATION

Do you love computers, networks, and systems? Do you want to know what it takes to be a technology whiz? Some web development and coding jobs require nothing more than a high school diploma or General Educational Development (GED) test. The GED test is a group of four tests in language arts, social studies, math, and science. Passing the GED is similar to receiving a high school diploma.

If the job requires you to work with a programming language, you may also need to obtain training in that language. Students can complete this training as part of a two-year program at a community college. They might also attend an accelerated coding program.

Many computer and technology jobs require a bachelor's degree. Website developers and digital designers usually need a bachelor's degree in computer science or programming. Software developers usually need a bachelor's degree in computer and information technology, engineering, or mathematics. Some developers may also need a master's degree in their chosen field. And programmers may need a bachelor's degree in CS.

Network and systems administrators as well as security analysts generally need a bachelor's degree in computer and information technology, engineering, mathematics, or a related field. Many jobs in these industries also require some form of certification. This is usually offered by specific vendors.

FINDING A JOB

Once you've met the requirements to work in your state, it's time to find a job!

JOB SEARCH TIPS

- Create a résumé that outlines your education and training as well as your work history. Even if you have not worked in the industry, highlight any relevant extracurricular experiences. These might include computer club or volunteer roles where you developed skills that can apply to the job you want. If you have an online portfolio, website, or social media account showcasing your work, include these links in your résumé.
- Keep in regular contact with the instructors and classmates from your training program. Ask them to put you in touch with others in the industry. The goal is to know many professionals in your industry. This can help you learn about possible job opportunities.
- Identify businesses, nonprofit organizations, or other establishments where you would like to work. Introduce yourself to the manager and ask for an informational interview. This can help you learn more about their business. Express your interest in working there, and leave them with a copy of your résumé.
- If you are invited to interview for a job, be prepared! Before the interview, research the business's history, mission statements, and offerings. Be ready to articulate why you want to work for the business and how you'd be a good fit for the role. Finally, have your own questions prepared for the interviewer. This shows you are thoughtful and serious about the job!
- Is there somewhere you really want to work, but there aren't any job openings for the role you're after? Consider applying for a different role. This can be an assistant, receptionist, or intern. This tactic is known as "getting your foot in the door." By doing this, you'll be in a strong position to be considered for your desired position when it becomes available.

LEVEL UP

Education doesn't stop once you land your first job. In many information technology jobs, workers must stay fluent in their programming language. This is especially important for computer programming and coding jobs. Employers may also require programmers and coders to learn new languages and techniques.

Even if additional training isn't required, it's important that technology experts seek out professional development opportunities. To obtain new and better roles, many tech professionals obtain certifications from specific vendors. Some vendors include Cisco, CompTIA, and Microsoft.

Tech experts also need to stay updated on the latest advancements in technology. This includes developments within their specific fields. This could mean taking courses and attending industry conferences. Some people might also gain inspiration by simply researching online, reading articles and books, and talking to other experts whenever possible.

Technology is constantly changing and improving. Keeping updated on new hardware, software, or security systems can help experts stay competitive in their chosen careers.

MEET A PRO: MEGAN SMITH, CHIEF TECHNOLOGY OFFICER

Megan Smith has long been passionate about computers, technology, and engineering. She studied at the Massachusetts Institute of Technology (MIT) from 1982 to 1988. At MIT, Smith designed, built, and drove a solar car in a race across Australia. She went on to work on innovative projects at Google. In 2014, President Obama appointed Smith to the position of US Chief Technology Officer (CTO). As the CTO, Smith focused on the ways information technology policy can shape the future of our nation.

In 2017, Smith became the chief executive officer of Shift7. Shift7 is a company that provides sales and marketing services to manufacturers and distributers. Smith is also the co-founder of the Malala Fund, where she contributes to girls' education around the globe.

Megan Smith speaking at a Take Action Day hosted by the Creative Artists Agency in 2017

TRADES AT WORK

EARNING POTENTIAL

The US Bureau of Labor Statistics provides estimated wage ranges for most workers in any given job category. The ranges below are from May 2023. While these estimates provide a sense of what you could expect to earn, actual salaries can vary greatly depending on where you work, your amount of experience, and any specialized skills you may have.

GROW YOUR POTENTIAL

No matter where your salary starts, there are various ways to grow your earning potential throughout your career. Here are a few ways you can boost your income while doing what you love.

Specialize in a specific type of technology. For example, you might decide to become an expert in a specialty programming language. Or perhaps you'll seek additional training in mobile app technology or artificial intelligence. In time, you may be able to charge more for your expertise.

Become your own boss. When working for an employer, you are typically limited to a set income, no matter how well the business does. But when you're self-employed, your earnings can increase with your experience and the success of your business. For some, self-employment means founding an independent software development company.

JOB CATEGORY	ANNUAL SALARY
Web Developer	$62,000-$116,000
Software Developer	$101,000-$168,000
Computer Programmer	$75,000-$130,000
Network and Computer Systems Administrator	$74,000-$122,000
Information Security Analyst	$90,000-$154,000

Another option that requires less of a financial investment might include providing information technology consultations remotely. Or you can provide web development and digital design services on a freelance basis.

Break into a new area of the industry. Consider specific clients who may need computer technology assistance. Perhaps you could build websites for recent college graduates. Or you could specialize in information security for local nonprofit organizations. You could even offer subcontract programming services to other computer programmers. If you get creative, you may discover multiple paths that branch from the one you're currently on.

FINANCIAL SMARTS

However you're making money, it's important to manage your finances wisely.

If you have an employer, you will receive a regular paycheck from them. This income will be your wages minus taxes. If your employer offers health insurance, retirement savings, or any other benefits, those will also be deducted from your take-home pay. Financial experts recommend you put about 20 percent of each paycheck into savings and try to keep an emergency fund with enough money to cover three to six months' worth of living expenses.

If you are self-employed, you will receive payments directly from your various clients. You'll need to track this income along with your business expenses, such as rent and supplies. Self-employed individuals must also pay their own taxes, generally four times a year, since they don't have an employer withholding taxes from each paycheck. Business owners use the remaining profits to pay themselves as well as fund savings accounts—for both themselves and the business!

If you have employees, you must pay both your employees and yourself. You must also manage your company's payroll, employee benefits, business insurance, and more. This takes a lot of work and organization. But hiring employees can bring many benefits, such as new skills and increased profits. It can also give you more time to focus on growing the business.

DO WHAT YOU LOVE!

Being a technology whiz requires commitment, dedication, and hard work. Success can require a knack for computers. It may also require plenty of practice with programming languages, hardware, and software. Still, many people in the computer and information technology industry find that the hard work is worth it.

Maybe you dream of becoming a programmer. Or you want to develop cutting-edge software or design unique websites. Perhaps you see yourself securing computer systems or troubleshooting network issues for big corporations. As long as you do what you love, you'll love what you do.

GLOSSARY

analyze–to examine something to find out what it is or what makes it work. Analytical relates to how something works. An analyst is a person who determines the meaning of something by breaking down its parts.

articulate–to clearly and effectively express oneself.

artificial intelligence–the ability of a computer to imitate intelligent human behavior.

assessment–the act of determining the importance or value of something.

breach–an act of breaking through.

compile–to collect or bring together items into one place.

component–one of the parts or units of a combination, mixture, or system.

configure–to set up or put together in a specific way.

cryptography–the writing and solving of codes or ciphers.

diagnose–to recognize something, such as a network issue, by signs, symptoms, or tests.

efficient–in a way that does not waste time or energy.

encryption–the process of turning data or text into a code or cipher to protect it.

Ethernet–a computer network that connects local area networks together with a physical cable.

exascale–a type of very powerful supercomputing.

fluent–able to use a language easily and accurately.

graphic–an image on the screen of a computer, TV, or other device.

innovative–marked by a new idea, method, or device.

integrate–to combine with or add to something else.

interchangeable–in place of each other.

manual–a book that contains instructions for how to use a machine.

output–data, power, or an electronic signal that comes out.

potential–capable of being or becoming, or something that could be.

server—a computer in a network that is used to provide services to other computers.

simultaneous—at the same time.

specialize—to develop expertise in a certain area, called a specialty. Specialized means suited to a particular purpose or occupation.

specification—a detailed description of how something should be designed or made.

subcontract—a binding agreement between someone who already has a binding agreement and another person who agrees to complete work from the first agreement.

technical—requiring special knowledge of a mechanical or scientific subject.

troubleshoot—to find and fix issues.

typography—the style, arrangement, or appearance of letters and text.

upgrade—an increase or improvement, or to make an increase or improvement.

vendor—a person or company that sells something.

World War II—from 1939 to 1945, fought in Europe, Asia, and Africa. Great Britain, France, the United States, the Soviet Union, and their allies were on one side. Germany, Italy, Japan, and their allies were on the other side.

ONLINE RESOURCES

Booklinks
NONFICTION NETWORK
FREE! ONLINE NONFICTION RESOURCES

To learn more about trades in technology, please visit **abdobooklinks.com** or scan this QR code. These links are routinely monitored and updated to provide the most current information available.

INDEX